ABOVE: *The author, Jill Davies, cropping thyme.*
COVER: *The Tudor Garden at the Tudor House Museum, Southampton.*

HERBS AND HERB GARDENS

Jill Davies

Shire Publications Ltd

CONTENTS

Introduction	3
Formal herb gardens	5
Creative herb gardening	12
The herbaceous border of herbs	16
The edible herb garden	20
Gardening advice	25
The dried herb and its uses	28
Suppliers of herb plants and seeds	30
Places to visit	31

Copyright © 1983 by Jill Davies. First published 1983; reprinted 1985, 1987, 1989, 1992. Shire Album 111. ISBN 0 85263 656 3.
All rights reserved. No part of this publication may be reproduced or transmitted in any form or by any means, electronic or mechanical, including photocopy, recording, or any information storage and retrieval system, without permission in writing from the publishers, Shire Publications Ltd, Cromwell House, Church Street, Princes Risborough, Aylesbury, Bucks HP17 9AJ, UK.

Set in 9 on 9pt Times and printed in Great Britain by C. I. Thomas & Sons (Haverfordwest) Ltd, Press Buildings, Merlins Bridge, Haverfordwest, Dyfed SA61 1XE.

ACKNOWLEDGEMENTS

Illustrations are acknowledged as follows: American Museum in Britain, Bath, page 3; French Government Tourist Office, page 23; Cadbury Lamb, pages 2, 4 (both), 7 (top), 12, 16; National Motor Museum Photographic Library, Beaulieu, page 19; National Museum of Wales (Welsh Folk Museum), page 15; Norfolk Lavender Ltd, page 27; Organic Farmers and Growers Ltd, page 26 (bottom); Lady Salisbury, page 7 (bottom); Southampton City Museums, front cover and page 5; Sussex Archaeological Society, page 29.

The drawings on pages 8, 9, 10, 20, 24, 25 and 26 are by D. R. Darton; those on pages 6 (bottom) 13, 14 and 18 are by Nicholas Driver.

The knot garden and, on the left, the lozenge-shaped herb beds at Pollock House, Glasgow.

The American herb garden at the American Museum in Britain, Claverton Manor, Bath (Reproduced by permission of the American Museum in Britain, Bath).

INTRODUCTION

Creating a herb garden of your own, of whatever kind appeals to you most, requires planning and forethought. This is best accumulated slowly and enjoyably long before your own garden has to be considered. It can be achieved by visiting National Trust gardens, friends' gardens, botanic gardens — anywhere that teaches you something about your own tastes and those of other gardeners.

During the summer months one can learn about colour, heights of plants and their conformations. Winter allows an uncluttered look at shapes, symmetrical or otherwise. It is a good idea to have a pocket book in which to record personal notes and memory joggers about these points. Even mistakes have their value, although it is more often the creator who sees the flaws in his own garden than the onlooker.

However, the key to success is often the degree of care and attention you devote to your plants. Look after them well and the garden will do the rest.

ABOVE: *A troy or turf maze at Wing, Leicestershire.*

LEFT: *The formal seventeenth-century garden at Hampton Court Palace.*

The Tudor Garden at the Tudor House Museum, Bugle Street, Southampton, was opened in 1982. It was designed by Sylvia Landsberg and executed by the Parks Department of the City of Southampton.

FORMAL HERB GARDENS

The shapes found in children's puzzles, games and Indian mandalas have strong links with the many patterns around us although their initial significance is often long forgotten. In a church in France there is a geometric pattern dating back to the fourth century AD. This was a precursor of the many garden mazes throughout the world, including the famous one at Hampton Court. The earliest example of a labyrinth is thought to be that at Knossos in Crete, celebrated in Greek legend. In the bronze and iron ages men created patterns with their complicated defensive circles of banks and ditches and what could be seen as forms of turf mazes. All these are very different from what we term 'gardening' today but they have influenced its development.

The shape of the knot garden is believed to be taken from beautiful Arabic designs which depicted elaborate and often geometric patterns (representations of the human form being forbidden by the Moslem religion). These were brought to Europe in the sixteenth century by sailors, mostly Dutch. Knot gardens were one of the earliest forms of an 'engineered' gardening layout, a conscious attempt to turn everyday plain plots of vegetables, herbs and flowers into an aesthetic design. These geometric shapes were laid down in the late fifteenth and sixteenth centuries, when they were fashionable both in Britain and in Europe.

The hallmark of a knot garden is the interflow and knotting effect of the shapes; it is a place that one can walk around in a leisurely and intimate way, admiring the contrasting tones, tints and textures of carefully placed plants. The knot garden within the Walled Garden at

The plan of the knot garden at Thornham.

Thornham Magna, Suffolk, is based on an original sixteenth-century design by Parkinson, a master herbalist, horticulturist, surgeon and astrologer who lived in Covent Garden, London. The British Museum has many knot garden designs drawn by herbalists of his time, but few have ever been put into effect. Those that were made have since been destroyed, either by the schemes of Capability Brown and other landscape designers of the eighteenth century or more recently in the two world wars.

At Hatfield House, Hertfordshire, Lady Salisbury has laid out a new knot garden which is full of scent and colour. Another one, 15 feet 6 inches (4.72 m) square, laid out by Sylvia Landsberg for Southampton City Council, is of an original design of 1582.

In time the parterre, which was fashionable in France, was introduced to Britain. Designed to be viewed from a castle, raised terrace or other vantage point, it was a more sophisticated and elaborate development. Spacious, formal and individual in concept, it differed from the knot garden by not following the flowing and continuous patterns of its predecessor.

However, although formality need not always mean intricacy, making a knot garden or parterre requires a great deal of time and space, but some simple designs have been prepared by the Herb Society and some very simple ideas were produced in the eighteenth century, an example being Queen Anne's Garden at Kew. Ideas can be drawn from floor tiles, weaving patterns, plate designs, cart wheels or fans; indeed almost any shape can be made by repeating or halving the original theme.

Parterre designs from the chapter 'The Ordering of the Garden for Pleasure' from Parkinson's herbal 'Paradisi in Sole'.

The knot garden at Hatfield House and, below, the plan of the garden, designed by Lady Salisbury and on which work began in 1980.

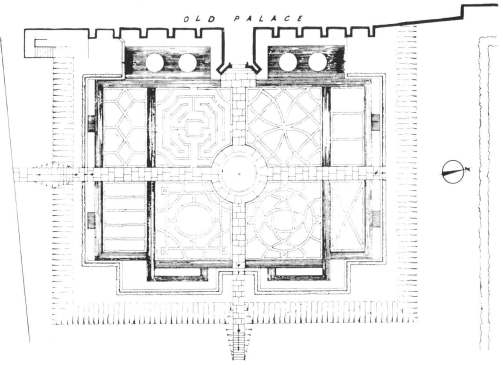

A design for a parterre.

HOW TO START

The ground must be level, even if the area around the design is not. So the first task is to dig and rake the site to an even tilth. Plastic templates are useful where a geometric shape has to be repeated a number of times, firstly using measuring tape and eye to obtain the initial outlines. For the continual repositioning of the templates, a grid laid out on the ground using small stakes and string is the most effective and efficient method. If the largest shape is cut first, the same piece of polythene can be used, merely cutting it down to form the smaller sizes. A black felt-tip pen (preferably water resistant) and large kitchen scissors are the ideal instruments for this.

With an uncomplicated design on a small scale, there is no need to mark out the ground with long-lasting outlines, especially if work is started in the morning and at least edged with a half-moon tool or even dug over by the end of the day. But for those with only odd moments or who want a more exclusive design, then a semi-permanent outline is needed. Sand trails, lasting only until wind or rain remove them, or small stones, tedious but effective, could be used, but the quick and efficient method is with lime; this remains for some weeks through most weather but can be rubbed out in the case of a drawing error. Using a bucket and trowel, sprinkle out slim lines of lime, which contrasts sharply with the dark soil.

Having dug over the soil within the designated shapes (soil health will be discussed in a later chapter), the next stage is to lay paths and some kind of permanent edging.

Paths. Gravel is perhaps the cheapest choice but it can scatter without a solid edge. A depth of 4 inches (100 mm) is ideal: laid straight on to the soil, this will allow for sinking in. Stone can be expensive and requires a good hardcore base covered with a 4 inch (100 mm) layer of sand. It must be made perfectly flat, using spirit levels, chisels and much wriggling and tapping. Bricks produce a very warm effect and look beautiful laid in unusual patterns, but this takes time and much effort.

Edges. See diagram of paths and edges.

Hedges. These provide a 'living' edge but are susceptible to frost damage and require constant clipping and keeping to an even height. Low ornamental hedges like box *(Buxus sempervirens)* and yew *(Taxus baccata)* are traditional, but other types include cotton lavender, myrtle, rosemary, rue, sage, wall germander and lavender. All are compact, enjoy a clipping and stay under 2 feet (610 mm) naturally.

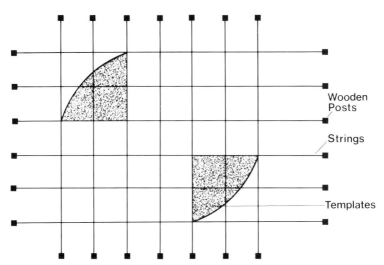

Marking out a geometric garden with templates. The templates are laid out on the string grid.

PLANTING

Planting should be the final event: the framework has been built up and now has to be filled in. If you are trying to cut costs the hedging plants could be bought and the remainder propagated.

In a knot garden it is exciting, although not traditional, to have startling height differences. Parterres, however, require much lower herbs.

Traditionally a knot garden had a centrepiece consisting of a large pot with a mature box or yew tree in it or even a simple sundial, but you could be adventurous, perhaps trying ornate topiary work. The height of the centrepiece helps to set off the low herbs.

Path designs: (a) Old slab path with lawn chamomile around the slabs; (b) a brick path with beds of creeping thyme at its edges; (c) loose chip path between treated timber edges with creeping comfrey at the sides; (d) a slab path between a low brick wall and an open bed.

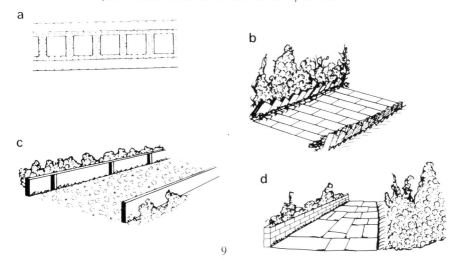

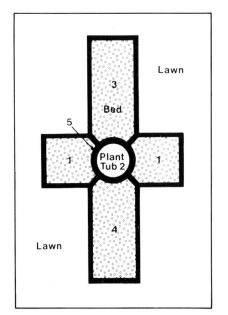
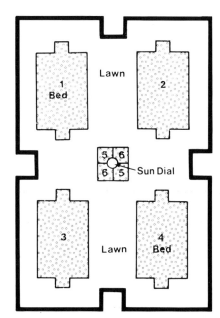

ABOVE: *Planting schemes. (Left) A simple cross: (1) lesser periwinkle (blue and low); (2) white horehound in the plant tub; (3) purple sage; (4) golden var. sage; (5) cotton lavender edges all round the path. (Right) A parterre diagram: (1) mace (yellow, mid height); (2) myrtle (Myrtus apiculata — evergreen, mid - low); (3) myrtle (Myrtus communis — evergreen, mid - low); (4) flax or linseed (blue, mid height); (5) lily of the valley (white, low); (6) marigold (yellow, low).*
BELOW: *The planting scheme for a fan-shaped bed, using 'simple' Labiatae herbs: (1) rosemary; (2) sage; (3) thyme; (4) marjoram; (5) dwarf lavender (as edging).*

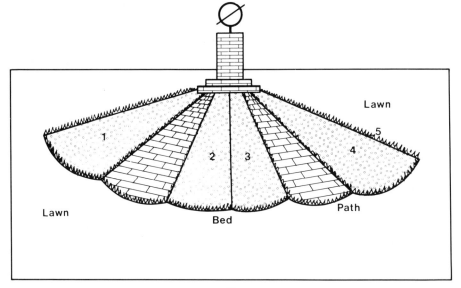

The planting scheme for the Thornham knot garden. (1) Hyssop (hyssopus officinalis). Perennial, height 60 cm; sunny situation. Flowers June to September, mostly blue, sometimes white and pink; can be used in stews or added dried to pot-pourri. (2) Rue (Ruta graveolens). Perennial, height 60 cm; sunny, well drained position. Flowers June to September, with greenish yellow heads. Bitter tasting herb, using for strewing in fifteenth and sixteenth centuries. (3). French sorrel (Rumex scutatus). Perennial, height 60 cm; rich soil and sun. Flowers July to August, greenish pink. Plant cuts copiously, lovely added to salad or soup. (4) Purple sage (Salvia officinalis var purpurascens). Perennial, height 30 cm; light soil in sun. Flowers July to September, purple like the leaves. Foliage lovely for flower and dish decorations. (5) Golden lemon balm (Melissa officinalis var aureum). Perennial, height 60-70 cm; rich soil, shelter, sun or shade. Flowers July to September, white while leaves are lemon smelling; ideal as herb tea, salads, flower arrangements. (6) Box (Buxus sempervirens). Perennial, height 10-30 cm if kept clipped; light soil and shelter. Inconspicuous flowers, evergreen leaves, useful dried in pot-pourri. Not marked on the plan but used as hedging to all beds, except where rue or purple sage are shown. (7) Bronze fennel (Foeniculum vulgare purpureum). Perennial, height 60-90 cm; rich soil, warm. Flowers July to September with large yellow umbrella heads. Leaves and seed useful for fish dishes. (8) White horehound (Marrubium vulgare). Perennial, height 30 - 60 cm; any soil but open situation. Flowers June to September in white whorls. Leaves good for tonic tea. (9) Alecost or costmary (Balsamita major). Perennial, height 60-90 cm; drained soil, sunny. Flowers July to August, daisy-like, yellow and white. Leaves in salads and soups, pepperminty smell early season. (10) Chamomile (Anthemis nobilis). Perennial, height 15 - 20 cm; sandy or loamy soil with sun. Flowers July to August, tiny white. Keep well clipped for lawn effect, use clippings, once dried, in pot-pourri. (11) Tansy (Tanacetum vulgare). Perennial, height 60 - 90 cm; sandy or loamy soil, sun. Flowers July to September, like flat yellow plates. Pungent smell of leaves, used as strewing herb in sixteenth century; sometimes flowers used in stews. (12) Sweet Cicely (Myrrhis odorata). Perennial, height 90-110 cm; drained, rich soil, sun and shade. Flowers May to June, white stars. Leaves provide natural sweetener to fruit. (13) Lily of the valley (Convallaria majalis). Perennial, height 15-20 cm, well drained and partial shade. Flowers May to June, white bells with sweet smell. Not for consumption. (14) Rosemary (Rosmarinus officinalis 'Jessop's Upright'). Perennial, height 90-110 cm; light soil, sheltered southerly position. Flowers April, May, mid-dark olive. Leaves used in cooking, teas, cosmetics and pot-pourri. (15) Cotton lavender (Santolina chamaecyparissus). Perennial, height 30-60 cm; drained, sunny situation. Flowers July, August, small yellow buttons. Aromatic silver, olive leaves (which should be kept clipped), ideal for pot-pourri. (16) Curry plant (Helichrysum italicum). Half-hardy perennial, height 60 cm; drained and sandy soil, sunny position. Flowers July to August with yellow 'everlasting' flowers. Silver-green leaves aromatic, ideal for pot-pourri. (17) Golden sage (Salvia officinalis var aurea). Perennial, height 30 - 35 cm; light, drained soil and sun. Flowers July to September, white and purple. Leaves attractive for decoration, can also be used in cooking; stews etc. (18) Golden thyme (Thymus aureum). Perennial, height 20-30 cm; light soil and sunny. Flowers June to September, pinky mauve. Leaves aromatic, useful in cooking and dried in pot-pourri for colour. (19) Wormwood (Artemesia absinthium). Perennial, height 90-140 cm; any soil, warm position. Flowers July to August, tiny greenish yellow clusters; striking silver - white foliage, lovely for flower arrangements. (20) Balsam poplar or balm of Gilead (Populus balsamifera). Perennial, height up to 5-10 metres; any soil and situation, useful for waterlogged situations to drain. Has sweet smelling buds, oil of which distilled for perfume and medicine. Flowers greeny-white, May to June.

The herb garden at Oakwell Hall and Country Park, Birstall, West Yorkshire.

CREATIVE HERB GARDENING

Herb gardens can be planned in 'free designs' such as are more usually associated with conventional gardening schemes. A water feature, be it stream, pond or lake, is a popular part of any garden, and a plan for a water-centred herb garden is illustrated. Water is often situated in partial shade; if it is not, shade can be created by planting water-loving trees, such as the alder and the poplar, which are both herbs that are very useful visually and medicinally. The sun-shy plants have been grouped below them in this design. Flowering interest through most of the season has been provided, and there are sharp differences in height, foliage, colour and texture.

Wild flowers make another unusual theme. A random arrangement is exciting, or one could enjoy the subtle variations of height and hue by making a group of yellow flowers, or red or blue. These can be grouped in circles, placing the tallest plant at the centre, working outwards to the shortest.

Even though annuals may have been sown to establish this garden they will seed freely and in a few years time the original arrangement will be unrecognisable with short and tall mixed up together — part of the magic of letting nature take over.

Enclosing the garden in an unusual and imaginative way is a challenge to many gardeners. Beth Chatto, at her Unusual Plant Nursery in Colchester, has used different varieties of that invaluable herb, the elder tree. Fast growing and easily trained to any shape with good thickness, they provide interesting foliage and delightful fruit in the autumn. At thornham one plant of each variety alternates with the common elder *(Sambucus nigra)* and the result is very striking.

Five varieties have been used. *Sambucus plumosa aurea* has cut-leaved golden foliage, with plumed filigree flowers in April and May. *Sambucus purpureum* resembles the common species but has purple leaves. *Sambucus variegata* has slender, pale green leaves with golden edges. *Sambucus lacinata* has finely divided mid-green foliage, which creates a dainty effect. *Sambucus aureum* has very golden leaves but is otherwise like the common species.

Southernwood or old man *(Artemisia abrotanum)* is a single plant or shrub attaining 90 to 130 centimetres (3 feet to 4 feet 3 inches) in height. It makes an excellent hedge and its aromatic foliage needs hard clipping to prevent it from straggling and breaking open. It thrives in full sun and on most soils, especially light sandy ones, and never fails to create attention. The copious clippings can be used for pot-pourri.

A planting scheme for a waterside bank. (1) Bistort (Polygonum bistorta). Perennial, up to 50 cm; moist to wet soil and shade. Flowers June to August, with beautiful pinky red spikes. Medicinal; leaves in salads. Best variety is 'superba'. (2) Meadowsweet (Filipendula ulmaria). Perennial, 90-120 cm; damp, waterside situations. Flowers June to September, creamy white. Highly medicinal or as a refreshing flower tea. (3) Common alder (Alnus glutinosa). Tree or shrub up to 2-3 metres. Flourishes beside, often in water. Bears beautiful green catkins in summer turning to attractive black in late summer/autumn. Medicinal. (4) Orris root (Iris florentina). Perennial up to 60 cm; best on light but moist soil enjoying either sun or partial shade. Flowers late May and June; attractive white with pale mauve and sometimes orange flower colours. Used as powdered 'base' for pot-pourri making. (5) Sweet violet Viola odorata). Perennial up to 15 cm; moist ground, shelter and partial shade. Flowers February to April, with tiny mauve faces. The leaves make a herb tea; flowers used for candying. (6) Variegated apple mint (Mentha suaveolens variegata). Perennial up to 20-30 cm; moist, rich soil and partial shade. Flowers, creamy mauve in August and September, but noted for beautiful white and green foliage, which remains striking in winter. Flower arrangers' delight. (7) Valerian (Valeriana officinalis). Perennial up to 60-90 cm; prefers medium damp soil out of direct sun. Flowers July to August with pretty filigree pinky white flowers set against dark green serrated leaves. Root used medicinally. (8) Greater periwinkle (Vinca major). Perennial forming ground cover mats up to 20 cm. Enjoys most situations but flourishes in damp and semi-shade. Flowers throughout summer, deep purplish blue. Whole plant used medicinally. (Some beautiful variegated varieties.) (9) Caper spurge (Euphorbia lathyris). Perennial 50-70 cm. Prefers deep shade and moist soil. Flowers greeny yellow in July. Attractive to flower arrangers; used medicinally, but milky sap dangerous to eyes. (10) Balsam poplar (Populus balsamifera). Perennial, height up to 5-10 metres. Any soil and situation, useful for waterlogged positions to drain. Sweet smelling buds, oil of which distilled for perfume and medicine. Flowers greeny-white May and June. (11) Sweet-scented Bergamot (Monarda didyma). Perennial up to 60-90 cm. Sunny position in a moist and rich soil. Flowers July to September with startling red flowers in whorls (some hybrids and species can be white, pink or purple). Flowers and leaves used as herb tea or dried for pot-pourri. (12) Lily of the valley (Convallaria majalis). Perennial 15-20 cm. Site on slope, as it does best on drained soil but loamy and in partial shade. Flowers May to July, white waxy bells. Medicinal and used in perfumery trade.

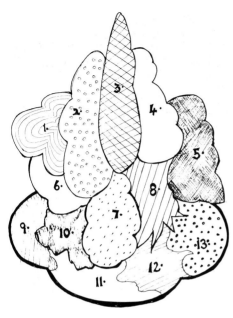

LEFT: *A planting scheme for a mixed wild flower garden. (1) Tutsan or sweet amber (hypericum androsaemum). Height 100 cm. Yellow flowers June to August and black berries in autumn. (2) Foxglove (Digitalis purpurea). Height 150 cm. Pinky purple flowers from June to September. (3) Teasel (Dipsacus fullonum). Height 200 cm. Pale pinky lilac flowers, green bracts from July to August. (4) Giant bellflower (Campanula latifolia). Height 100cm. Pale blue flowers from July to August. (5) Danewort or dwarf elder (Sambucus ebulus). Height 95 cm. Pinky white flowers July to August with autumn fruit. (6) Soapwort (Saponaria officinalis). Height 80 cm. Pale pink flowers July to September. (7) Columbine (Aquilegia vulgaris). Height 70 cm. Deep blue mauve flowers from May to June. (8) White campion (Silene alba). Height 80 cm. White flowers May to August. (9) Winter aconite (Eranthis hyemalis). Height 10-15 cm. Mid yellow flowers from January to March. (10) Astrantia (Astrantia major). Height 60 cm. Pinky white flowers from May to July. (11) Pyramidal bugle (Ajuga pyramidalis). Height 15 cm. Dark blue flowers from April to May. (12) Dwarf spurge (Euphorbia exigua). Height 25 cm. Mostly green bracts and flowers June to October. (13) Red clover (Trifolium pratense). Height 30 cm. Deep red flowers May to September.*

A planting scheme for a yellow wild flower circle. (1) Mullein (Vervascum thapsus). Mid yellow flowers, June to October. (2) Large-flowered evening primrose (Oenothera erythrosepala). Magnificent yellow flowers, July to October. (3) Ox-eye daisy (Chrysanthemum leucanthemum). Flowers white-petalled but with large deep yellow centres, May to August. (4) Marigold (Calendula officinalis). Deep yellow/orange flowers, June to November. (5) Coltsfoot (Tussilago farfara). Mid yellow flowers March and April, followed by leaves. (6) Melilot (Melilotus officinalis). Mid yellow flowers, June to September. (7) Arnica (Arnica montana). Deep bronze/yellow flowers, June to August. (8) Cowlsip (Primula veris). Mid/pale yellow flowers, April and May. (9) Dandelion (Taraxacum officinalis). Mid yellow flowers, almost all the year; ideal plant. (10) Primrose (Primula vulgaris). Pale yellow flowers, March to May.

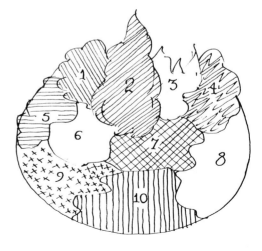

The herb garden at St. Fagans Castle, near Cardiff.

Two herbs which will create much interest in the garden are thorn-apple *(Datura stramonium)* and sea-holly *(Eryngium maritimum)*. As an annual, thorn-apple grows rapidly from being a seedling in May to reach a height of 1 to 1.25 metres (3 feet 3 inches to 4 feet) by July and August. It has a branching habit and produces large white trumpet-shaped flowers in June, July and August. These develop into the most bizarre green seed pods, covered in large spines. When ripe, the pods burst open and scatter numerous deadly, dark brown seeds to the ground. The whole plant is highly poisonous and aromatically narcotic, but it is a native to Britain as well as many other parts of the world. Plant thorn-apple in masses, give it plenty of room and enjoy its eccentricities but keep children away from it.

Although spiny, the sea-holly is a less alarming plant. Its roots are highly valuable to the green pharmacist, and the young flowering shoots can be eaten like asparagus, being sweet and pungent. There are many spectacular varieties of *Eryngium* but the sea-holly is a familiar sight on the coast, where it remains a rather scrubby and stout plant, keeping its head low from salt and wind. In the garden, given rich loamy, part sandy soil, it can reach 60 to 80 centimetres (2 feet to 2 feet 6 inches). A perennial, it has a deeply penetrating root system and produces beautiful, blue.green, egg-shaped flowers in July, while the spiny foliage, mostly green and glaucous, echoes the blue-green tinge nearer the top of the plant. Plant sea-holly in the garden with other silver foliage herbs and the results will be most striking. Cut, dried or steeped in glycerine, it is an unusual delight to the winter flower arranger.

The herb garden at Knebworth House, Hertfordshire, is based upon the designs of Gertrude Jekyll.

THE HERBACEOUS BORDER OF HERBS

A herbaceous border that is simple to make and easy to maintain may seem to be a contradiction in terms and it certainly would have been in the eighteenth century when this type of garden originated. Even nowadays, when hybridisation, dwarfing hormones and other human interventions have produced more free-standing plants, some tying and staking is necessary to provide such a glorious show. But this is not so with herbs. They are amongst the most robust of plants, natural survivors, always evolving with their surroundings rather than facing extinction. They stand without artificial support. It is only unnaturally bred high-performance plants with unbalanced flowerheads and weakened stems that require a gardener's constant helping hand.

A herbaceous border is rather like a shop window display, with many different things to be taken in at a glance. Tall plants should be at the back, short ones at the front and continuing interest throughout the seasons should be provided.

The depth of the border must always reflect the scale of the garden, so it is difficult to suggest a figure, but one should always aim to fill what has been dug — better busy ground than empty spaces for weeds and work. A path or a sweep of lawn in front is essential, to give a clean-cut appearance and emphasise the fullness of the border.

One of the duties of the herbaceous border is to provide a constant supply of plants for flower arranging, cooking, the medicine cupboard and pot-pourri. This should be achieved without spoiling the appearance of the border and incorporated into an effortless gardening ven-

The author, Jill Davies, cropping lovage.

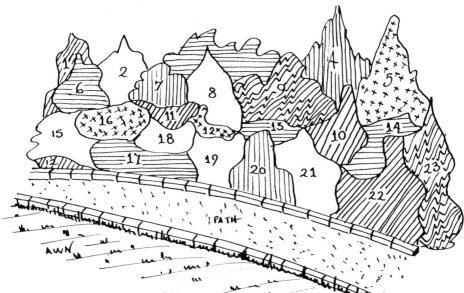

A planting scheme for a herbaceous border. All the plants are perennial. Plants 1-5 are tallest, 6-10 are mid height, 11-14 are ground cover, 'carpeting' height and placed around stepping stones, for ease of movement; 15-23 are medium low. (1) Golden var elder (Sambucus aureum); height 180-450 cm. (2) Witch hazel (Hamaemelis virginiana); flowers yellow in autumn; height up to 360 cm. (3) Sweet briar (Rosa rubiginosa); flowers pink, June and July; red hips, autumn; height 200 to 300 cm. (4) Gravel root (Eupatorium purpureum); flowers August to October, cream, purple; height 150-300 cm. (5) Tree mallow (Lavatera arborea); flowers July to October, pinky lilac; height 200-400 cm. (6) Bear's breech (Acanthus mollis); flowers August to November, white, lilac on spikes; height 75 to 200 cm. (7) Elecampane (Inula helenium); flowers July to August, bright yellow; height 120-150 cm. (8) Rosemary 'Seven Seas' (Rosmarinus 'Seven-Seas'); flowers April to May, pale blue; height 100-120 cm. (9) Deadly nightshade (Atropa belladonna); flowers September to October, purple; black fruit after; height 50-200 cm. (10) Male fern (Dryopteris filix-mas); green fronds, brown under spores all year; height 70-150 cm. (11) Creeping comfrey (Symphytum repens); dark green leaves, white flowers, spring to Autumn; height 20-30 cm. (12) Caraway thyme (Thymus herba-barona); flowers June, pale pink; height 5-12 cm. (13) Creeping evening primrose (Oenothera missouriensis); flowers midsummer, yellow; height 10-30 cm. (14) Creeping catnep (Nepeta cataria 'Beth Chatto'); flowers all summer, blue; height 10-20 cm. (15) Common sage (Salvia officinalis); flowers July to September, purple white; height 45-50 cm. (16) Sweet bay (Laurus nobilis); flowers May to June; evergreen; height 45-60 cm (bushy). (17) Saffron crocus (Crocus sativus); flowers autumn, lilac purple; height 10 cm. (18) Cotton lavender (Santolina chamaecyparissus); flowers July, yellow; height 30-50 cm. (19) Purple plantain (Plantago major var purpureum); flowers summer, leaves purple; height 20-30 cm. (20) Lavender 'Dwarf Hidcote' (Lavendula 'Dwarf Hidcote', flowers June to August, purple, blue; height 25-35 cm. (21) Golden marjoram (Origanum majorana var aurea); flowers July to September, white and small; height 25-30 cm. (22) Pasque flower (Anemone pulsatilla); flowers spring, blue, violet; height 5-30 cm. (23) Ladies' mantle (Alchemilla vulgaris); flowers spring to mid autumn, green, yellow; height 10-50 cm.

ture. The simple shape and layout with the use of only perennials and copious ground cover plants reduce the labour enormously. Also, if a deep border is chosen (or even a shallow one), stepping stones or log-rounds placed amongst the plants facilitate its attendance, eliminating the need to put on heavy shoes and subsequently to hoe over the footprints.

The choice of plants in the scheme illustrated provides for basic culinary requirements (rosemary, sage and thyme). There are also several medicinal herbs and a host of attractive and unusual ones combining shape and colour. So as well as being useful the arrangement is visually unpredictable and yet well blended.

The common counterpart of the purple plantain, *Plantago major,* has extensive medicinal uses and is unequalled as a 'first aid' herb for stings and bites. This purple variety (propagated by Beth Chatto) has larger leaves and flower spikes and its beauty and boldness are accentuated if it is placed amongst foliage of silver, grey and white.

Witch-hazel is a herb most people are familiar with — in a bottle as a remedy for strains and bruises. Few realise what a beautiful shrub it is, with yellow star-shaped flowers appearing quite unexpectedly after leaf fall in the autumn.

The male fern is more striking than some of its relatives and, although useful to a trained medical herbalist, it can be nighly dangerous to the curious. In spring it forms an attractive crown shape with tight fronds which unfurl to become green and lanceloate, bearing green and then brown spores on the underside, from summer through to autumn.

The tree mallow is a shrub which has a long and prolific flowering period. Pinky lilac in colour, the flowers are similar to but larger than those of its relative the marshmallow *(Althea officinalis),* one of the kindest and safest medicinal aids. Pruning by a quarter to a half in spring encourages the prolific show of colour throughout summer and autumn.

Thought must be given to the grouping of these plants although if you follow the plan provided an attractive arrangement will come about; a great deal of its initial success (time eventually always fills and mellows) depends on the choice of individual plants at the nursery, whether bushy, tall, healthy or whatever. Plants often need a mass effect, especially in a large border, so grouping in threes or fives is often necessary to avoid a pinched look. Plants that are vigorous and large to begin with must be left to their own glory, but otherwise the gardener should be bold.

The herb garden designed by Moyra Burnett and planted in the ruins of Beaulieu Abbey. (The Abbey is only open to visitors on a general admission ticket to Beaulieu).

THE EDIBLE HERB GARDEN

One of the most important aspects of growing herbs is the culinary delights they produce. Whether you with to add the merest flavour to a dish or to impart a dominant taste, it is best always to have the fresh herb for when dried it will lose much of its quality and vitality. Happily, there are herbs that can be picked and used fresh all the year round and with a little help from a cloche or belljar many more can be kept in production over the winter.

This is the most functional kind of herb garden and it is important for it to be within easy reach of the house. So planning should begin from the kitchen door or back door. A path, if not already laid, is advisable and you should make this as prominent a feature as the garden itself. One of the cheapest and safest underfoot materials is peashingle, which affords no smooth surface for ice to make slippery. Edged with brick or even with a soft-leaved plant such as southernwood or lavender, the path can sweep into the culinary herb garden itself, bringing unity and ease of plucking.

The shapes for the garden can be as formal or free flowing as desired, but good use of stepping stones and internal paths is advisable so that it is an invitingly functional area and not just beautiful.

Here is a basic list of plants, all of which have specific but simple requirements. Each of the two sections, culinary herbs and salad herbs, is split into four groups corresponding to the seasons of the year.

CULINARY HERBS

SPRING
Sweet Cicely *(Myrrhis odorata)*. February to November. Pick flowers, leaves and even stem for use as a natural sweetener of puddings. Perennial, 90 to 120 cm.

Ideas for herb garden designs: (a) converging paths bisect the beds; (b) alternate beds and slabs line a brick path; (c) raised beds of wedge formation, contained within low brick walls, enable the scent of the herbs to be better appreciated.

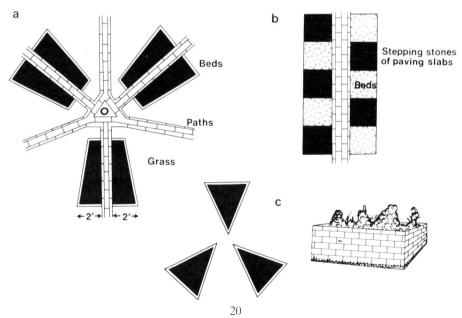

Enjoys drained, loamy soil. Propagation by seed.

Welsh or green onion *(Allium fistulosum)*. April to October. Pick stem tops for good salad base or addition to stews etc; onion flavour. For maximum foliage prevent flowering. Perennial, 30 cm. Enjoys light soil. Propagation by seed in spring.

Chervil *(Chaerophyllum sativum)*. April to July. Pick leaves before flowering; ideal in soups, as garnish or fresh in salads. Has a sweet, clean taste. Annual, occasionally biennial, 40 cm.

SUMMER

Dill *(Anethum graveolens)*. July to August. Pick leaves and, later, seed. Leaves used in any fish dishes, soups, sauces; likewise seed, which is also used in pickling. Aniseed flavour, similar to fennel. Annual, 60 to 90 cm. Enjoys drained light soil and sun. Propagation by seed in March under cover, or outside in late April.

Pot marjoram *(Origanum onites)*. July to October. Pick leaves and flowers and use in stews, soups or stuffings, as a garnish and fresh in salads. Imparts a slightly lemony flavour, giving a dish a fresh lift, lovely with heavy meats. Perennial, 60 cm. Enjoys light drained soil and warmth. Propagation by seed or cuttings.

French tarragon *(Artemisia dracunculus)*. June to October. Pick leaves (a little goes a long way). Use in soups, stews, sparingly in salads, for sauces and tarragon vinegar. Has aniseed taste, numbing the tongue for a while if of good quality. Perennial, 60 to 90 cm. Will grow in most soils, but needs shelter, especially over the winter. Propagation by cold frame cuttings in spring. Replace plants every three years to ensure taste quality.

AUTUMN

Lemon thyme *(Thymus x citriodorus)*. Most of the year for picking, especially if sheltered over winter. Flowers for garnish and leaves and flowers for stuffings, stews, *bouquet garni;* lovely added to tomato soup. This lemon variety is the best cooking one. Half-hardy perennial, 30 cm. Drained, sunny position. Propagation by seed, spring cuttings or layering side shoots. The taste of lemon dwindles after three years, so it needs to be replaced. Prune to keep non-woody.

Common sage *(Salvia officinalis)*. Spring to November. Pick leaves and flowers for stuffings, stews, soups. Imparts traditional flavour, lovely in sausages. Common in East Anglia. Perennial and evergreen, 45 to 60 cm. Enjoys most soils except very heavy ones; sunny situation. Propagation by seed or easily from cuttings in May and June. Replace every three to five years for best flavour. Prune to keep non-woody.

Garden mint *(Mentha x spicata)*. Spring to November. Pick leaves for mint sauce, or as addition to new peas and potatoes, soups, garnishing. Perennial, 30 to 45 cm. Prefers moist soil and partial shade. Propagation by plant's stolons in autumn; it becomes rampant if not contained by planting in sunken old bucket.

WINTER

Winter savoury *(Satureia montana)*. All year. Pick leaves and flowers for fish dishes, stuffings, soups, *bouquet garni.* Pleasant gentle taste. Perennial, 25 to 45 cm. Enjoys light soil, chalky if possible, and full sun. Propagation by seed in September, or divide roots in spring or autumn, or take spring cuttings.

Rosemary 'Jessup's Upright' *(Rosmarinus officinalis* var J.U.). Pick leaves all year round and use flowers from April to June in soups, stews, roast lamb and potatoes. Strong taste, so use sparingly. Perennial, evergreen, 90 to 150 cm. This variety grows upright and enjoys cropping. Light, sandy soil; sheltered southerly site. Propagation, best from cuttings in spring or autumn.

Sweet bay *(Laurus nobilis)*. All the year. Pick leaves for soups and stews, especially good with mince; and put with basil. Strong taste, so use sparingly. Perennial evergreen, 90 to 300 cm. Enjoys medium light soil, sun and shelter; may need protection in hard winter. Propagation, best from heel cuttings or layering lower branches in July to August.

SALAD HERBS

SPRING

Chives *(Allium schoenoprasum)*. March to September. Pick the leaves and flowers

all summer for decoration of salad, while the leaves form a good salad base. Cutting encourages growth. Faint onion flavour. Perennial, 15 to 45 cm; forms clumps. Thrives on rich soil, grows on any, prefers half shade. Propagation, by seed initially, afterwards divide clumps; keep flowers trimmed for better growth.
Cowslip *(Primula veris)*. April to June for the bright yellow flowers which provide pretty garnish for spring salads. The leaves impart a tangy flavour, similar to watercress. Perennial, 20 to 30 cm. Likes woodland soil conditions (friable soil and leaf mould), with sunny situation. Propagation by seed in spring or root division in autumn.
Coltsfoot *(Tussilago farfara)*. February to March for the flowers, which provide pretty deep yellow garnish for salads; nutty flavour, excellent as a cough remedy. The leaves, which appear after flowering, are good chopped on omelettes. Perennial, 15 to 40 cm. Enjoys most soil, shade or sun. Propagation by seed in spring or division of copiously produced plants in autumn.

SUMMER
Sweet basil *(Ocymum basilicum)*. July to September or October. If lucky, by cutting the leaves two crops can be obtained in a season. Used fresh in salads or cold mayonnaise (sparingly), it imparts an unusual but strong flavour. Lovely made into vinegar for salad dressings. Half-hardy annual, 30 to 40 cm. Prefers drained soil, plenty of sun and shelter. Propagation: sow seed indoors in April, put out late May or June.
Green and gold purslane. *(Portulacca oleracea, Portulacca var sativa)*. June to September. The green variety provides yellow flowers and the gold one crimson or purple flowers, beautiful as a garnish. The leaves are refreshing and tangy as a salad base. Annual, 10 to 15 cm. Enjoys sandy soil and sunny position. Propagation: sow seed indoors in April.
Borage *(Borago officinalis)*. June to September for the leaves, best eaten when young and pale green. The white flowers, with bright blue stars, in July add a delightful garnish to salad. Annual, sometimes biennial, 60 to 90 cm. Enjoys any drained soil in the sun. Propagation by seed in spring, but it self-seeds very freely.

AUTUMN
Nasturtium *(Tropaeolum majus)*. June to November for leaves which are lovely in sandwiches or salads, imparting a hot, tangy flavour, while the flowers (July to October) are red, yellow or beige, providing a garnish with good flavour. Annual, 30 to 180 cm, creeping growth. Will grow anywhere; likes sun. Propagation by seed in March or April; sets seed freely itself once established.
Marigold *(Calendula officinalis)*. June to November for orange petals, lovely whole or chopped into salads, adding a nutty flavour. Gives colour to an egg mayonnaise. Annual, 30 to 40 cm. Any soil, but loves full sun. Propagation by seed in March or April, but sets seed freely once established.
Curly parsley *(Petroselinum crispum)*. June to November for leaves, which if cloche-covered will last all winter. Rich in vitamins A, C and E, iron and other minerals; should be a daily addition to any salad. Light tangy taste. Remove flowers to promote leaf growth. Biennial, 15 to 20 cm. Excels in moist rich soil, sun and partial shade. Propagation, by seed in February; thin out in May.

WINTER
Corn salad or lamb's lettuce *(Valerianella locusta)*. Pick leaves from October to following spring, when it seeds. This fresh winter salad is delightful and generally does not need protection. Annual, 10 to 20 cm. Any soil in an open situation. Propagation by seed in autumn or early spring.
French sorrel *(Rumex scutatus)*. All year round; overwinters well, sometimes needs cloche protection. Has tangy lemony flavour, beautiful in salads and cold sauces, especially with prawns and crab. Perennial, 60 cm. Good soil and sun. Propagation by seed in spring, or divide roots in spring or autumn.
Salad burnet *(Poterium sanguisorba)*. Leaves all year round; flowers May to August for pretty red garnish. Leaves have sharp nutty flavour. Perennial, favours chalky soil but grows in most; likes a sunny situation. Propagation, by

The extensive ornamental gardens of Villandry in France.

seed in April.

Some of the first ornamental gardens consisted entirely of herbs and vegetables and a magnificent example is the Italian-designed sixteenth-century garden at Villandry in France. It still flourishes, tended by monks.

It is possible to embellish your own edible herb garden with some vegetables. A simple edging of spinach is attractive or the white-stemmed, dark-green leafed Swiss chard makes a good alternative to spinach, with the added advantages of overwintering and being perennial. There are some ornamental varieties of vegetables, like the pink, white and lilac flushed cabbages with serrated edges (seeds from Thiompson and Morgan), but these are not for eating.

Herbs and vegetables have long been companions and are known to produce better flavours if placed in certain partnerships: for instance, summer savoury improves the taste of French beans, both in the soil and in the pot. There are many such favourable combinations and such plants often seek their own neighbours, wherever they are placed.

GARDENING ADVICE

At the outset of any garden landscaping, thought must be given to the young plants taking root while the beds where they are to be planted are being prepared. However, if one starts constructional work in the winter, then little can be done about this. Even if the plants for the outlines and focal points, such as trees and hedges, are to be purchased for quick effect, then probably the infills will have to be from seed, cuttings and root division. But do not attempt this without making sure that source material is free of disease, robust, true to type and correctly labelled.

SEED

Seed is an often easy and unusually prolific means of propagation, but it is important to buy seed raised in Britain — a more difficult task than one might imagine. Local seed produces greater initial success and makes for a mature plant, more able to stand up to disease and cold winters, while seed raised in different climes is not genetically equipped to do this. Seed packaging instructions are generally good, so any particular needs of the plant will be mentioned. Otherwise the method remains the same for all. Using trays on the windowsill or in the greenhouse is convenient but if you possess cold frames or cloches then use these. A 'seedling' or 'universal' compost is advisable as these are steam sterilised, preventing weed competition. Using a tray, fill it to quarter full and then firm down evenly, using a block of wood. Sprinkle the seed on carefully, cupping the seed in the palm of one hand and pinching small quantities between the forefinger and thumb of the other hand. Keep an equal distance between the seeds, however large or small. Now cover the seed but sieve the compost first to give the seedling every chance to reach the surface with ease. Large seeds need deep soil coverage, while small ones require just a dusting. Label them clearly with pencil or indelible ink pen, and whether they are placed inside or outside, make sure the temperature is retained at an even level, taking day and night differences into consideration.

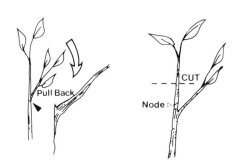

(Left) A heel cutting; (right) a nodal cutting.

CUTTINGS

Cuttings can be taken as early as February, right through to late September. Choose cuttings from new but firm growth, which has sap rising but is not flimsy; for this reason March to April growth is best. The *slip* or *heel* cutting is the one most likely to root because it has the greatest amount of rooting hormones situated at the point where it is taken. The *nodal* cutting has a slightly smaller concentration of hormones at the cutting point.

The ideal length of cutting is 4 inches (100 mm) but it does vary according to the material available. Foliage should be carefully removed to save the plant's energy for root production rather than for feeding unnecessary leaves. Always pinch off foliage, taking care to avoid tearing the stem, as this would create an open wound where infection might enter. However, the top one or two leaves should be left to provide light and oxygen for the plant. Next dip the cuttings in rooting hormone if you feel that nature needs help. Knock off any excess hormone powder to avoid callousing (the formation of a hard scab, inpenetrable to young pushing roots). Use graded hormone powders for soft, semi-hard and hard-wood cuttings as they are more sensitive and obtain better results.

Insert the cutting to half its length, having first cut a slit in the compost with a knife. Water well with a fine spray, label and leave it, curbing the desire to see if it has rooted. Subsequently water it often but in small quantities; under watering causes stress, and over watering mildew, rot and disease.

Tough, unpampered rooting conditions produce tough, healthy plants, so choose a simple site to pitch your cuttings. The shelter of a southerly wall, cold frame or cloche is ideal. Sophistications like heating cables are not necessary and may produce quick yet sickly plants. Either make your own soil of, say, two parts sharp sand, two parts peat and one part loam (this will require hand weeding because it is not sterilised) or buy some 'universal' compost. The small amount of loam addition in both these composts means that potting up can be slightly delayed without harm to the plant, as there is some food present.

POTTING UP

Well rooted cuttings or firm stout seedlings need a better home with more food if they are to avoid becoming weak and sickly (etiolated). After potting up place them in a shady but warm situation. Shade allows the roots to firm up without draining the plant's system, although too much shade results in damp and stagnation. Preferably put them on level sand, which retains water well and is an ideal medium into which the plant can root if it is left longer than intended.

ROOT DIVISION AND SIDE SHOOTS

These are two of the easiest methods of propagation. Root division means choosing a large perennial plant like a clump of bergamot or lemon balm and, using a sharp spade, chopping it into two or four. One piece is left and the other one or three are taken to be replanted instantly in a large hole lined with peat to encourage quick root establishment.

Side shoots can be encouraged by the addition of peat around the stem base in spring or autumn. 'Irishmen's cuttings' result and plants easily produced in this way are thyme, marjoram, cotton lavender, sage and many more.

PREPARATION OF THE GROUND

When making the herb garden of your choice, bear in mind that creating is fun but that large herb gardens, and even small and intricate ones, need care and attention. Experience teaches you to cultivate only what you can look after.

Perennial weeds should be removed, otherwise they will be a constant nuisance to anything that is planted. Double digging goes to the depth of the most deep-rooted weeds and encourages aeration of the soil for earthworm activity, letting in sunshine and oxygen and unlocking nutrients for the newly planted searching roots.

Afterwards, levelling can be carried out, generally by raking in several directions, removing the large stones but leaving the smaller ones for drainage.

Herbs are happy in most soil types but an open, aerated soil will suit them all, even without the specific considerations of chalk or sand, which can easily be added if necessary. Humus keeps all soils in a lively balanced state and the addition of well rotted compost is ideal.

Soil should contain four main ele-

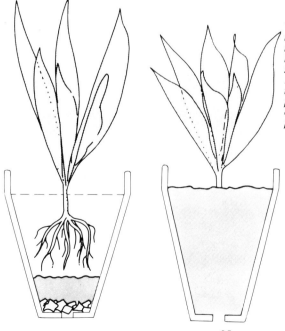

When potting up (left), ensure there is plenty of room for the roots and do not cram them in but let them 'dangle'. Keep the soil level on the stem the same from seed tray to pot. (Right). Keep the soil level half an inch (12 mm) or more below the top of the pot to allow for watering room without spillage and loss of soil through puddles.

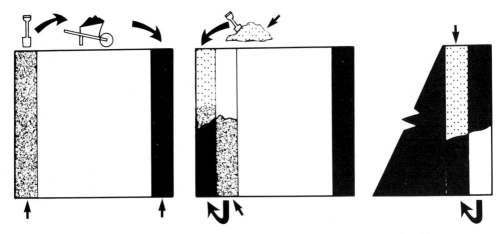

Double digging. Move the soil from the first trench (on the left) to the end of the bed. Add compost to the trench, then fill it with soil from a second trench. Repeat for a third and continue until you reach the end. The last trench is filled with the soil that came from the first trench.

ments, which are often lost or used up and so need replacement, for they may not be found in humus alone.

They are lime, nitrogen, potash and phosphates. *Lime* breaks down sticky wet clay soils and unlocks the soil to release other plant foods. There is no value in excessive amounts: generally 13.5 kilograms (30 pounds) of fine powder to 25 square metres (270 square feet) is sufficient, hydrate of lime being the best form. *Nitrogen* produces growth, soft tissues, leaves and shoots and is found in sulphate of ammonia. Half a kilogram (1 pound) is sufficient for 25 square metres (270 square feet). Feeds of macerated comfrey are an alternative source. *Potash* gives quality, colour and flavour. Apply 2 kilograms (4½ pounds) of sulphate of potash to 25 square metres (270 square feet). *Phosphates* build up hard fibrous tissue, encourage root action and bring plants to maturity, seed and fruit production. Use superphosphate at 1.5 kilograms (3 pounds) per square metre (10 square feet). All these additions should be applied in late February.

PATHS AND MARKING OUT

Paths are essential and should be given first consideration in muddy winter conditions.

How to lay out shapes using string, marker posts and liming lines has already been explained.

TOOLS

The right tools make a great difference to the enjoyment of one's gardening.

A 'Real' hoe has a pivoting head which digs deeper into the soil.

Carefully choose a spade and fork, correct weight and length being important, while secateurs, knives and scythes should be kept clean and oiled so they work to maximum efficiency.

Rotavators may seem the quick and easy answer to arduous digging, but they need constant attention and can be difficult for a weaker-framed person to control. Overused, they cultivate the soil to a point where the particles are so finely ground that a clay consistency develops, often resulting in a 'pan' or area of impregnably compacted soil, causing drainage problems. Occasionally used, they can be useful, but are best hired for a specific job.

Similar machines, but man-powered, are the various hand-pushed ploughs. Jalo makes a good one with various fitments for hoeing and weeding and cultivating, all to varying depths and widths, but you need stamina to use it.

A good lawnmower to suit the particular needs of the grass is essential, as well kept lawns enhance beautiful flower beds.

The most useful tool of all is the hoe and the one illustrated does not cause backache or the desire to turn to chemicals in order to fight the constant battle against weeds.

The herb gardens at Norfolk Lavender Ltd, Heacham, Norfolk.

THE DRIED HERB AND ITS USES

GATHERING THE FRESH HERB

Although using the fresh herb is usually preferable sometimes this is not convenient or desirable. Summer-grown annuals must be harvested for winter use and a pot-pourri will take months of picking and storage as the flowers bloom, being gathered at their most perfect state for the final combination. A herb tea could take even longer if not just the leaf and flower but also the hip or berry are included.

Harvesting should always take place after the dew has gone but before the morning has become too hot. Choose healthy, clean material, picking for quality, but do not completely ravage the plant. Everything should be collected when at the peak of condition, before there is any loss or release of its energies. Leaves, therefore, are collected as they unfurl or just after unfurling; flowers are harvested in bud or when just opened, so that the aromatic oils have not escaped. Berries should be collected at colour peak (for example, rosehips when bright red), while roots should be dug up when the top foliage is dormant, either in spring or preferably in autumn before the plant has used up some of its goodness for winter fuel. If bark needs to be collected this should be done when the sap rises in the spring; it is much easier to peel it away with a knife when this occurs, taking care not to ring-bark the tree, but always lifting in patches.

DRYING AND STORAGE

Drying herbs must be done with care for the quality of the finished product depends on it. Having picked clean, healthy specimens only, lay them on a cake cooling rack (preferably a wooden one) in a warm spot, out of direct sunlight to avoid loss of flower and colour. The temperature should be above 22C (77F) but not exceed 35C (95F). Cover with muslin and check the herbs, turning them daily until they are properly dried out. Leaves take about three days, but the bulkier the herb or the part of it being dried is then the longer the process takes. So watch and wait, discarding anything slightly damp which might stagnate and cause mildew. Roots are best chopped and may take three weeks even after this.

A condensation test is advisable to establish thorough drying. For this, place the herb in a glass jar and tightly seal the cap, placing it in a warm position. If water droplets appear after some hours, then the water content is still too high, so resume the drying process. When the herbs pass this test place them in dark jars with airtight lids, labelling them clearly. Kilner jars are excellent for storage but being clear glass they must be kept in a dark place, as light destroys colour and medicinal quality.

Though often convenient, freezing is not a good way to store herbs, because the vitamin content is destroyed and one might as well enjoy the full food value even of a garnish. However, some commercially frozen herbs are so well produced that they retain some goodness, so buy these, rather than attempt your own. The Herba brand label is a fine example.

A HERB PILLOW

Some dried herbs can be put to use as soon as they are ready, for instance in a herb pillow. Here is an excellent recipe for a herb pillow to help one fall asleep at night: two parts skullcap (leaf), two parts elderflower (flowers), half part hops (flowers, very bulky), four parts catnep (leaves and flowers).

A POT-POURRI RECIPE

Collection for pot-pourri can start with the first flowers of spring, such as catkins, going on to horse-chestnut flowers and rose petals and ending with winter broom. What to include is a personal choice but here is an extremely simple pot-pourri recipe:

1 part catkins
½ part rosemary (leaves)
½ part rose petals (bud)
1 part rose hips
1 part horse-chestnut (flowers)
½ part ground ivy (leaves)
1 part borage (flowers)
2 parts orris root powder (dry fixative)

A few drops of essential oils like geranium, rose otto and peppermint. Put more oil in every four months or so, to revive the fragrance.

HERB TEAS OR TISANES

Herb teas and tisanes are one of the most immediate ways of imbibing herbs for pleasure and medicinal value. A tea is an infusion of a dried herb in hot water to release its flavour and healing qualities. When drunk, it goes straight into the bloodstream and acts more quickly than pills or capsules.

Here is one example of a tea, called 'China Light, tisane, which both tastes beautiful, smells refreshing and helps to cure headaches: mix lavender, rosemary and a small quantity of hops, according to taste, and add a few drops of peppermint oil. Use one teaspoonful per mug of boiling water. Infuse in a single teapot with a lid, for ten to twenty minutes. Sweeten with honey, if desired.

Dried herbs should play a large part in winter cookery although a great many herbs are available fresh. Dried thyme not only adds a delightful flavour to many dishes but provides the protection of an antiseptic. Rosemary tastes good on roast potatoes and also enlivens the circulation, an important benefit in the colder months. Sage serves mainly as an ingredient in stuffings but also helps to heal coughs and other chesty ailments so common in winter. And what could be more delightful than unscrewing a jar of perfectly dried, bright blue borage flowers, to decorate a salad on Christmas Day?

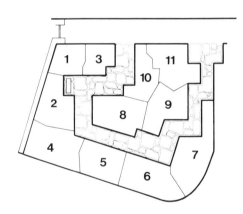

The Physic Garden opened in 1981 at Michelham Priory, East Sussex. The plants are arranged in groups according to their medicinal use, as follows: (1) rheumatism, gout and painful joints — mugwort, mustard and lily of the valley; (2) use in the household — sage, soapwort and tansy; (3) childbirth and children's diseases — stitchwort, horehound and violet; (4) heart, lung and blood disorders — plantain, mullein and coltsfoot; (5) wounds and broken bones — self-heal, comfrey and salad burnet; (6) bites, stings, burns and poisons — orris, mallow and calamint; (7) digestion, stomach and liver — sorrel, chicory and rue; (8) depression and dreams — poppy, feverfew and borage; (9) head, hair and skin — fumitory, love-in-a-mist and pennyroyal; (10) animal husbandry — balm, sweet Cicely and chickweed; (11) eyes, ears and teeth — marigold, wall germander and bistort.

SUPPLIERS OF HERB PLANTS AND SEEDS

SEED MERCHANTS
John Chambers, 15 Westleigh Road, Barton Seagrave, Kettering, Northamptonshire NN15 5AJ.
Chiltern Seeds, Bortree Stile, Ulverston, Cumbria LA12 7PB.
Emorsgate Seeds, Emorsgate, Terrington St Clement, King's Lynn, Norfolk PE34 4NY.
Nature Conservancy Council, PO Box 6, Huntingdon, Cambridgeshire PE18 6BU. (For wild flower seed.)
The Seed Bank, 44 Albion Road, Sutton, Surrey. (For herb and wild flower seed.)
Suffolk Herbs Ltd, Sawyers Farm, Little Cornard, Sudbury, Suffolk. Telephone: 0787 227247.

HERB FARMS AND NURSERIES
Arne Herbs, The Old Tavern, Compton Dundon, Somerton, Somerset TA11 6PP. Telephone: 0458 42347.
Askett Nurseries, Flowery Field, Aylesbury, Buckinghamshire HP17 9LY. Telephone: 0899 279635.
Barwinnock Herbs, Barrhill, Ayrshire. Telephone: 046582 338.
Beth Chatto, Unusual Plant Nursery, White Barn, Elmstead, Colchester, Essex.
Bruisyard Vineyard Herb Gardens Church Road, Bruisyard, Saxmundham, Suffolk IP17 2EF. Telephone: 072875 .
The Cottage Herbery, Mill House, Boraston, Tenbury Wells, Worcestershire WR15 8LZ. Telephone: 058479 575.
Cranborne Garden Centre, Cranborne, Wimborne, Dorset. Telephone: 07254 248.
Devon Herbs, Thorn Cottage, Burn Lane, Brentor, Tavistock, Devon PL19 0ND. Telephone: 0822 810285.
Dorwest Herb Growers, Shipton Gorge, Bridport, Dorset. Telephone: 0308 897272.
Elly Hill Herbs, Elly Hill House, Darlington, Co Durham DL1 3JF. Telephone: 0325 464682.
Clive Essame, 1 Beaumont Cottages, Gittisham, Honiton, Devon EX14 0AG. Telephone: 0404 850815.
Fallowfield Herbs, Morrells Farm, Whatlington, East Sussex. Telephone: 0424 870387.
Foliage Scented and Herb Plants, Walton Poor Cottage, Ranmore, Dorking, Surrey. Telephone: 04865 2273/4731.
Gardiners Herbs, 35 Victoria Road, Mortlake, London SW14 8EX. Telephone: 081-878 7981.
Hatfield House, Hatfield, Hertfordshire. Telephone: 07072 62823.
Heaven Scent Herbs, Rose Cottage, Heathfield, Langford Budville, Wellington, Somerset TA21 0RP. Telephone: 0823 400689.
Herb and Heather Centre, West Haddlesey, Selby, North Yorkshire YO8 8QA. Telephone: 0757 228279.
The Herb Farm, Peppard Road, Sonning Common, Reading RG4 9NJ. Telephone: 0734 724220.
The Herb Garden, Hall View Cottage, Hardstoft, Pilsley, Chesterfield, Derbyshire. Telephone: 0246 854268.
The Herb Nursery, Thistleton, Oakham, Rutland LE15 7RE. Telephone: 0572 83658.
The Herbary, Prickwillow, Ely, Cambridgeshire CB7 4SJ. Telephone: 035388 456.
Herbs in Stock, Whites Hill, Stock, Ingatestone, Essex. Telephone: 0277 841130.
Hexham Herbs, Chesters Walled Garden, Chollerford, Hexham, Northumberland. Telephone: 0434 681483.
Highland Herb Nursery (Department H), Newton of Petty Dalcross, Inverness IV1 2JQ.
Hill Farm Herbs, Park Walk, Brigstock, Northamptonshire NN14 3HH. Telephone: 0536 373694.
Hollington Nurseries Ltd, Woolton Hill, Newbury, Berkshire. Telephone: 0635 253908.
Iden Croft Herbs, Frittenden Road, Staplehurst, Kent TN12 0DN. Telephone: 0580 891432.
Judy's Country Garden, Louth Road, South Somercotes, Louth, Lincolnshire. Telephone: 0507 358487.
Lathbury Park Herbs, Lathbury Park, Newport Pagnell, Buckinghamshire. Telephone: 0908 610316.

Lomond Herb Garden, Lomond, Horsehill, Hookwood, Horley, Surrey. Telephone: 0293 862318.
Lower Severalls Herb Nursery, Lower Severalls, Crewkerne, Somerset TA18 7NX. Telephone: 0460 73234.
Manor House Herbs, Wadeford, Chard, Somerset.
Marle Place Plants, Marle Place, Brenchley, Kent TN12 7HS. Telephone: 089272 2304.
Netherfield Herbs, The Thatched Cottage, 37 Nether Street, Rougham, Bury St Edmunds, Suffolk IP30 9LW. Telephone: 0359 70452.
Norfolk Herbs, Mill Farm, Wendling, Dereham, Norfolk NR19 2LY. Telephone: 036287 211.
Norfolk Lavender Ltd, Caley Mill, Heacham, King's Lynn, Norfolk. Telephone: 0485 70384.
The Old Manse Herb Garden, The Old Manse, Bridge of Marnoch, Huntly, Aberdeenshire AB5 5RS. Telephone: 04665 873.
The Old Rectory Herb Garden, Rectory Lane, Ightham, Sevenoaks, Kent. Telephone: 0732 882608.
Old Semeil Herb Garden, Strathdon, Aberdeenshire. Telephone: 09756 51343.
Parkinson Herbs, Barras Moor Farm, Perranarworthal, Truro, Cornwall TR3 7PE. Telephone: 0872 864380.
Pepper Alley Herbs, The Spice Warehouse, Pepper Alley, Banbury, Oxfordshire OX16 8JB. Telephone: 0295 253888.
Samares Herbs à Plenty, Samares Manor, St Clement, Jersey, Channel Islands. Telephone: 0534 79635.
Sellett Hall Herb Garden, Whittington, Kirkby Lonsdale, Via Carnforth, Lancashire LA6 2QF. Telephone: 05242 71865.
Selsey Herb and Goat Farm, Waterlane, Selsey, Stroud, Gloucestershire GL5 5LW. Telephone: 0453 766632.
Selsey Herb Shop, 22 Castle Street, Cirencester, Gloucestershire.
Southwick Country Herbs, Southwick Farm, Nomansland, Tiverton, Devon EX16 8NW. Telephone: 0884 861099.
Suffolk Herbs, Sawyers Farm, Little Cornard, Sudbury, Suffolk CO10 0NY. Telephone: 0787 227247.
Valeswood Herb Farm, Little Ness, Shrewsbury, Shropshire. Telephone: 0939 260376.
Wells and Winter Ltd, Mereworth, Maidstone, Kent.
Westhall Herbs, Church Lane, Westhall, Halesworth, Suffolk IP19 8NU. Telephone: 050279 646.
Woodbine Cottage Herbs, Back Bank, Whaplode Drove, Spalding, Lincolnshire PE12 0TT. Telephone: 0406 330693.

TOOLS
The 'Real' hoe is available from Organic Farmers and Growers Ltd, 9 Station Approach, Needham Market, Suffolk. Telephone: 0449 720838.

COURSES
Osho School for Herbalists and Natural Healers, with the College of Herbs and Natural Healing (EC recognised), 2 Bridge Farm Cottage, Station Road, Pulham Market, Diss, Norfolk IP12 4SJ. Herbalism and Natural Healing, with Herb Identification in the Wild (Jill Davies, Kitty Campion and Sensai Cayenne).

USEFUL ADDRESSES
Henry Doubleday, HDRA, Covent Lane, Bocking, Braintree, Essex CM7 6RW.
The Fields Studies Council, Preston Montford, Montford Bridge, Shrewsbury, Shropshire SY4 1HW. (Courses in plant and herb identification.)
The Herb Society, PO Box 599, London SW11 4RW. (Publishes quarterly magazine.)
The Natural Therapeutics Centre, Suryodaya, Gislingham, Eye, Suffolk IP23 8JG. Telephone: 0379 783527.
Royal Botanic Gardens, Kew, Richmond, Surrey TW9 3HB. Telephone: 081-940 1171.

PLACES TO VISIT

Intending visitors are advised to check hours and dates of opening before making a special journey. In many of the places listed the herb garden comprises only a small part of the grounds open to the public and separate admission to the herb garden alone is not normally available.

Abbey Dore Court Garden, Abbeydore, Hereford. Telephone: 0981 240419.
Abbey House Museum, Kirkstall, Leeds. Telephone: 0532 755821.
Acorn Bank, Temple Sowerby, Cumbria. Telephone: 09663 3883.
Beaulieu Palace House and Gardens, Beaulieu, Brockenhurst, Hampshire SO4 7ZN. Telephone: 0590 612345.
Bewdley Museum, The Shambles, Load Street, Bewdley, Worcestershire DY12 2AE.
Cambridge University Botanic Garden, Cambridge. Telephone: 0223 350101.
Castle Drogo, near Chagford, Newton Abbot, Devon. Telephone: 064743 3306.
Chelsea Physic Garden, 66 Royal Hospital Road, London SW3. Telephone: 071-352 5646.
Chenies Manor House, Chenies, Buckinghamshire. Telephone: 02404 2888.
The Cloister Garden, Peterborough Cathedral.
Coke's Cottage, West Burton, Near Pulborough, West Sussex.
Craft at the Suffolk Barn, Fornham Road Farm, Great Barton, Bury St Edmunds, Suffolk IP31 2SG. Telephone: 028487 317.
Curtis Museum, High Street, Alton, Hampshire. Telephone: 0420 82802.
Dartington Hall, near Totnes, Devon. Telephone: 0803 862224.
East Lambrook Manor, South Petherton, Somerset. Telephone: 0460 40328.
Emmanuel College, St Andrew's Street, Cambridge. Telephone: 0223 65411.
Fishbourne Roman Palace, Salthill Road, Fishbourne, Chichester, West Sussex. Telephone: 0243 785859.
Gaulden Manor, Tolland, Lydeard St Lawrence, Taunton, Somerset. Telephone: 09847 213.
Guernsey Herb Garden, Ashcroft Hotel, Sous L'Eglise, St Saviour's, Guernsey, Channel Islands. Telephone: 0481 63862.
Gunby Hall, Burgh-le-Marsh, Skegness, Lincolnshire. Telephone: 075485 212.
Hatfield House, Hertfordshire. Telephone: 0707 262823.
Herschel House and Museum, 19 New King Street, Bath, Avon BA1 2BL. Telephone: 0225 336228.
Hever Castle, Edenbridge, Kent. Telephone: 0732 862205.
Hidcote Manor Garden, Hidcote Bartrim, Chipping Campden, Gloucestershire. Telephone: 0386 438333.
Highbury, West Moors, Wimborne, Dorset. Telephone: 0202 874372.
Kew Gardens, Kew, Richmond, Surrey. Telephone: 081-940 1171.
Knebworth House, Knebworth, Hertfordshire. Telephone: 0438 812661.
Liverpool University Botanic Gardens, Ness, South Wirral, Cheshire. Telephone: 051-336 2135.
Michelham Priory, Upper Dicker, Hailsham, East Sussex. Telephone: 0323 844224.
Norton Priory Walled Garden, Tudor Road, Runcorn, Cheshire WA7 1SX. Telephone: 0928 569895.
Oxford Botanic Garden, High Street, Oxford. Telephone: 0865 276920.
Scotney Castle Garden, Lamberhurst, Tunbridge Wells, Kent. Telephone: 0892 890651.
Sissinghurst Castle Garden, Sissinghurst, Cranbrook, Kent. Telephone: 0580 712850.
Stoneacre, Stoneacre Lane, Otham, Maidstone, Kent.
Threave Garden, Castle Douglas, Dumfries and Galloway. Telephone: 0556 2575.
Tradescant Trust, St Mary's Church, Lambeth, London SE1. Telephone: 071-261 1891.
Tudor House Museum, Bugle Street, Southampton, Hampshire. Telephone: 0703 24216.
Welsh Folk Museum, St Fagans Castle, Cardiff CF5 6XB. Telephone: 0222 569441.
Westbury Court Garden, Westbury-on-Severn, Gloucestershire. Telephone: 045276 461.
West Green House, Hartley Wintney, Basingstoke, Hampshire.
Westminster Abbey College Gardens, Westminster, London SW1. Telephone: 071-222 5152.
Wisley Garden (Royal Horticultural Society), Wisley, Woking, Surrey. Telephone: 0483 224234.